ANOTHER TIME

By

Rodney.C.Terrell

ISBN: 1-4033-7212-8 (e-book)
ISBN: 1-4033-7213-6 (Paperback)

This book is printed on acid free paper.

1stBooks – rev. 10/30/02

FORWARD

Due to a phenomenon that happened to me in my youth for over fifty years I struggled to understand how it could be possible. In time I not only found an logical answer to the phenomenon but my investigation lead me to understand why we Humans are so different from the other species and other phenomenon's that have baffled Man through the ages. I might have called the book "The Relativity of Probabilities" instead of Another Time for it is based on the relativity of history, scientific facts, theories and the many religious beliefs. It led me to the understanding of a Supreme Being not in outer space or in the imagination but in the future in the form a Super Human Race with powers beyond our comprehension. Part of the book is fiction the other part is fact and theory.

For al those like myself that have encounter an unexplained phenomenon I hope this book will bring some comfort.

Rodney C. Terrell

CHAPTER 1

Within the hour Mark would know if his force field shield worked. If the shield failed Mark along with all the others abroad the Time Craft would be blown to bits and scattered through out time and space.

It all began one night when a Galaxy suddenly disappeared while being photographed by the astronomer Westwood. He had seen Galaxy disappeared but before but never so sudden. The next morning Westwood contacted his astrophysicist friend Clarence and told him of the event. Can you explain how this could be possible? How long did it take to disappear? It was gone in just the blink of the eye. Only a Black Hole could cause it to disappear that quick; however it would have to be a very large Black Hole to swallow up a Galaxy. You know Clarence for a long time

1

I've been thinking What if there is a Black Hole large enough to swallow up the whole Universe. You forget Westwood such a Black Hole would not explain why the Galaxies were moving away from one another. I have a theory how it could be possible. I remember when I was 10 years old my father took me to a science museum that had an exhibit where you would spin one ball after another around a large funnel shape drum. As I spun the balls I noticed as they moved toward the smaller hole at the bottom of the drum they distanced themselves from one another in much the same way a Black Hole works. You may be right Westwood I'm going to present your theory to my colleague.

Clarence and his astrophysicist colleague spent the next several months in searching for anything that would support Westwood's Black Hole theory. Eventual they turned up a couple of things that confirmed the theory. One: 90% of matter the Universe should contain were missing. Being drawn into a Black Hole was a logical explanation. Two: Galaxies are speeding up the further they move away from their origin. This means some kind of force was effecting them. The only known force that great would have to be a Super Black Hole. The question was how could such a Black Hole come into existence.

Being an astronomer Westwood had watched colliding Galaxies and wondered what caused them to collide. After giving it much thought Westwood figured the logical explanation was the further the Galaxies traveled away from the force of the Big Bang that created them the weaker that force became. Then their gravity pulled them together creating a colliding Galaxy. As the stars of the colliding

Galaxy died they created a Super Black Hole. Westwood's

Super Black Hole theory was eventually proven.

As Diagram 3 And 3A Shows The Doppler Shift Can Be Deceptive. It Is Possible The Galaxies Are Being Pulled Into A Super Black Hole As Shown In The Diagrams. Eveident Of This Would Be If Older Galaxies Were Increasing Their Speed Away From The Point Of The Big Bang.

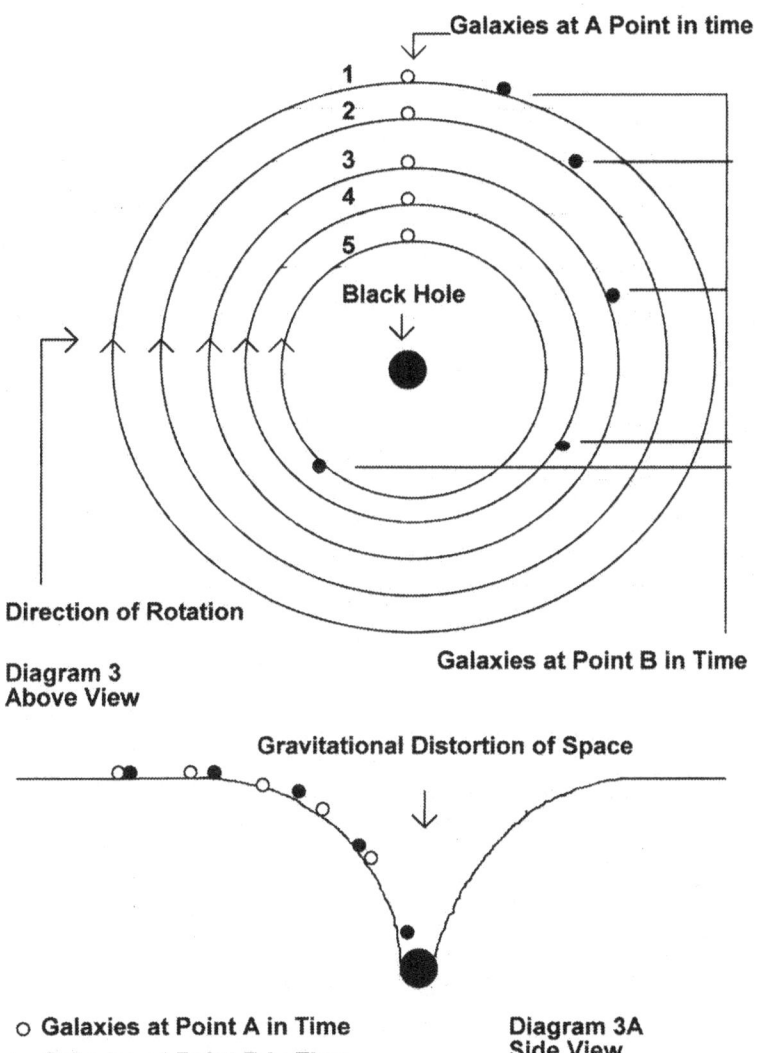

Galaxies at A Point in time

1
2
3
4
5

Black Hole

Direction of Rotation

Diagram 3
Above View

Galaxies at Point B in Time

Gravitational Distortion of Space

o Galaxies at Point A in Time
● Galaxies at Point B in Time

Diagram 3A
Side View

CHAPTER 2

The worlds leading scientists meeting at the United Nation concluded there was not time enough left to build the number of space ships needed to carry all the people safely away from the Black Hole destine to destroy the Earth. This left them with only one choice, they had to find a way to travel back in time; thus was born Project Escape. The idea behind the project was to find a force great enough to propel a craft beyond the speed of light thus make traveling back in time possible. The only thing that could supply such force was the gravitational force of a Super Back Hole.

For many years the United States searched for a way to use antigravity as a method to propel a space ship but the project was unsuccessful. Five years after the famous physicist Houser took charge of the project the antigravity drive became operational. Since the crew and passengers of an antigravity craft would be in a none gravitational environment they would not be effect by sudden acceleration and course change at tremendous speed. The main concern was colliding with space object. Beyond the speed of light the collision with even a pea size object would turn such a craft and its occupants into a cloud of dust.

The space ship chosen to install the antigravity drive was the United Nations space ship Alpha. Built for an expedition to the Andromeda Galaxy it was the only space ship large enough to accommodate the ten thousand people needed to set up a colony back in time. Fear among the people they would not be able to get aboard one of the space ships caused no shortage of volunteers for the first trip aboard the untested Time Craft. Volunteers for the trip poured in from all over the Earth as well as the Moon and Mars colonies.

Now just minutes until departure if all went well the Earth would disappear in the blink of an eye as the Time Craft soared into space. Sitting there among the other crew members going through the countdown; Mark's thoughts drifted back to when he was a child and first became aware of the force destine to destroy his world. He wondered what it would have been like growing up in a world where one could look forward to a carefree life and old age. However; he would never know such a life for at an early age he became conscious of the fact that his struggle for continued existence and that of Mankind was the same. Although many of his friends had committed suicide rather than live in a dying world. He decided however to do all within his power to assure the continued existence of the Human Race so his parent's deaths as well as the deaths of all those other billions of human beings that lived and died in the past would not have been in vain. It was this desire that sustained him when the development of his force field shield seemed doomed to failure.

Now at departure time panic griped Mark, what if his shield failed! Then the Captain engaged the antigravity drive and the Alpha soared up through the Earth's atmosphere into space. As the speed neared the speed of light the force field shield gave off a blue glow from being struck by microscopic particles. On occasion however there would be a bright flash off the shield caused by harmless objects too small to require a course change. So far so good Mark thought; but the real test would come once they exceeded the speed of light. When the Alpha passed the speed of light the blue glow turned to blackness. At the moment of blackness a shout went up from the crew as they rushed over to congratulate Mark on the success of his force field shield. The success of the shield may Mark feel like a huge weight had been lifted from his shoulders. Passing the speed of light the Alpha's rate of acceleration increased immensely and shortly reached the half way point at which time the Captain reversed its antigravity drive to slow it down. This was a crucial time, for even a second could throw them off course by a thousand years.

CHAPTER 3

The plan was to travel back in time 20 thousand years. That would place them in the era when modern man's direct ancestor Cro-Magnon Man roamed the Earth. The place chosen to establish the Time Colony was Egypt. Since the indication was that Cro-Magnon man mostly roamed Europe it was unlikely there would be an encounter between them thus changing history.

As the Alpha's speed slowed below the speed of light the blue glow reappeared then faded from the viewing screen and Mark found himself gazing down on an Earth that he had never known except from ancient video discs. Wherever he looked there was lush green vegetation unlike the barren Earth with its domed cities and farms of his time. Once the Alpha reached its destination the Captain put it in a hovering mode a mile above the Earth. He then ordered two of the Alpha's scout crafts to gather information about this Earth.

Mark thought what a strange sight it must be for the inhabitants below looking up at the three miles across saucer shaped Time Craft with two of its saucer shaped scout crafts streaking across their sky. Within hours the crafts had canvassed the Earth from the North to South Pole examining every thing from the different species to the composition of its soil. By the time the scout crafts returned to the Alpha it became apparent there was no Cro-Magnons. Was it possible the computer had malfunctioned and they had not arrived at their destination? A star check confirmed their calculations were correct. Now the question was should they land and establish a Time Colony as planed or do further exploration. The decision was made to send two scout crafts to the historical sites of Neanderthal Man and Cro-Magnon Man remains. Each scout craft would carried a party of five made up of a Captain, archaeologist, a crew of two, and a robot to do the digging. Scout craft number 1 was sent to Dusseldorf Germany. Dusseldorf was not far from where history had shown the remains of a Neanderthal Man were found. Number 2 Scout craft headed to Monte Carlo near the location of Cro-Magnon Man remains. Mark took charge of the number 1 scout craft. Finding the locations were no problem since the total history of the Earth was in the Alpha's computer.

Slowly Marks scout craft came to rest on its tripod. Looking out of the window the assault on his senses from all the different life forms was so overwhelming it was difficult for him to get his mind back to the job of locating the remains of Neanderthal Man. Mark ordered the robot ahead to the dig site as a precaution against the unknown. Once the robot reached the dig site the archaeologist on leaving the scout craft to oversee the digging simply vanished. Mark ordered the robot back to look for the archaeologist. Using his VCS (video-communicator-scanner) he contacted Captain Ray White of scout crafts number two. "Ray: Mark here; don't let anyone go outside." Thanks; but you are to late we had a bit of a rough landing so one of the crew went outside to check and see if there was any damage and he disappeared. Yes, I just lost my archaeologist. Ray stay on the VCS with me while I contact Captain Walker and inform him of what has happened. A robot's picture appeared on Marks hand held VCS screen beside Captain White. Communication robot C1 here what can I do for you? C1; this is urgent! Contact Captain Walker and put him on. Within seconds Captain Walkers face appeared replacing the image of the robot. Walker here, how is it going down there. We have a problem, Ray lost a crew member and I lost my archaeologist. How? I don't know. It seems a step outside the scout craft is a step into oblivion. How is that? You simply disappear. What! We can't live out side of the scout craft. Dose it effect the robot? No, the robot seems to be functioning O.K. Then have the robot locate the missing archaeologist and crew member. We already tried and they were unable to detect any sign of them. Well, since the robots are operational carry on with the mission just using the robots. I will get in touch with the science department and see if they have an answer to the disappearance of those men. I'll see you when you get back.

Mark instructed the robot to return to the dig site. Within the hour the robot reported there were no remains of

Neanderthal Man so he ordered it back to the craft. As the scout craft soared up into the sky on its return to the Alpha Mark wondered if the other party had been successful in finding the remains of Cro-Magnon Man. He would not have to wonder long for as his craft approached its docking bay under the Alpha he noticed the other scout craft was also docking. He realized if they had found the remains of Cro-Magnon Man they would not have returned so soon. On disembarking Mark boarded a shuttle car waiting for him. It whisked him off to an elevator that took him up to Captain Walker's office on the fifth level.

On entering the office a receptionist ushered him into the conference room where he took a seated at an oblong table. Among the others at the meeting was Professor Scott head of scientific research, members of the Planning Committee, and Ray White the Captain of the number 2 scout craft.

Captain Walker rose and addressed the group. I know this is going to sound impossible gentlemen but the expedition showed there are no remains of Cro-Magnon or Neanderthal. It seems they didn't become extinct as history recorded; the search showed they never existed. As the members of the Planning Committee sat in stunned silence Captain Walker continued; gentlemen this fact leaves us with a paradox. We don't exist; or we shouldn't. This is one of the reasons I have called this meeting. We have to find out how it is possible for us to be alive when there is no evident of our ancestor ever existing.

Now for the other reason I've call the meeting. To go outside the of Alpha or its scout crafts mean death. I have asked Professor Scott of the scientific research department for an explanation of this phenomenon. Scott: Gentlemen; when we exceeded the speed of light we ceased to exist in real time; that is the time in which we were born. We now exist in what I call Our Time. That Cro-Magnon and Neanderthal existed in history but not in Our Time tell me that somehow it's up to us to see that they do. We were so afraid we would alter history it never occurred to us that our very existence depended on changing it. Captain Walker: If what you say is true then we don't exist? To anyone in real time we don't. Let me explain; when we exceeded the speed of light we were all in the same environment as the Alpha and as long as we stay in that environment to each other we are normal. "However; to leave the Alpha means death. I am sorry I didn't think of this possibility before we lost the archaeologist and the crew member. Gentlemen; for now we are prisoners in our time traveling environment. After the death of the archaeologist and crew member I did an experiment. I put a fly from the laboratory in a small metal cage attached to a wire and lowered it outside the Alpha. When I retrieved the cage all that remained of the fly was the inorganic matter that made up its body. It is lucky the Alpha has a metal hull other wise we all would be dead. As Captain; this new knowledge is going to force us to change our Time Colonizing plan. Since none of man's ancestors exist at this time I suggest we travel back in time and see if we can find Australopithecus (southern ape) the species thought to be the first pro-human. All in favor raise your hand. We have nine for and two against. Representative Quaker and Burke may I ask why you are against the plan. Quaker: I'm not against the plan; I just think the crew and passengers should have a vote on it. I see; however as Captain of the Alpha I am not only responsible for the craft but also the passengers. Until we reach our destination I will continue to carry out my job. When we reach our destination then the crew and passengers will have the right to vote on whatever they wish. What is it about the plan that you disagree with Mr. Burke? You have answered one of my questions but I have a question for Professor Scott. What if a person wanted to return to his or her real time; would they still be prisoners of the Alpha? If they returned to the time they left then they should not be effected however if a person returned before or after the time in which they left they would be faced with the same problem we have. Captain Walker: Are there any more questions? No! Good let us be on our way.

CHAPTER 4

As the Alpha headed further back in time, Mark thought how ironic it was that his generation would not only be responsible for saving the Human Race from extinction but also for its very existence. This was just another paradox of time travel. After the Alpha landed Captain Walker ordered robot guards stationed around its outer perimeters.

There were three type of robots the R25, R50 and R99 the 99 stood for a robot 99% Human with the missing 1% being emotion. With the strength of fifty men the robots made excellent guards. Its eyes used to give it mobility was also a TV camera that sent pictures back to the central control computer aboard the Alpha. By monitoring the pictures and the robot's four other sensory systems; touch, taste, smell, and hearing, the central computer could send a directive to the robot on what action to take under any given circumstance.

Once the Alpha was made secure; robots were sent to search for the human ancestor ape Australopithecus and soon returned with a large male. However after testing it's DNA the conclusion was that it could not be man's ancestor so the decision was to alter the DNA of a few dozens. Afterwards they traveled forward in time to see if altering the DNA of Australopithecus would bring about the existence of Cro-Magnon Man.

The location chosen to land was fifty miles from what was to become Dusseldorf Germany; where history had shown Cro-Magnon Man had existed. As the Alpha approached the landing sight Mark noticed a group of fifteen Neanderthals that included six women and three children. He was glad that at least the DNA experiment had worked up to the Neanderthals. Unlike a spacecraft that slowly appears as it descends from space a Time Craft just suddenly appears. When the Neanderthals looked up and saw the 3-mile across Alpha hovering above them at first stood frozen in their tracks unable to decide what to do next. Looking up at the object that suddenly appeared above them they must have thought it was going to crush them. Suddenly they began running in all directions. Mark was glad they did otherwise they would have been crushed by one of the Alpha's city block size tripod limbs. After the Alpha landed escalators that ran up and down each of the tripod limbs was turned on. The escalator was necessary for at rest on its tripod the bottom level was still three hundred and fifty feet above the surface. Beside the escalators the Alpha also had in its center a huge telescoping tube containing a freight elevator that extended from the surface up through it's ten levels.

After the Alpha had set motionless for a while the Neanderthals curiosity overcame their fear and they slowly and cautiously approached one of the tripod limbs. The leader fascinated with the escalator's motion put one foot on it lost his balance and fell. Jumping to his feet he and the others ran toward a nearby forest. However, once among the trees they realized they were still beneath the monster that came down from the sky. Exhausted from running they slowed downed to a walk and continued toward the outer edge of the Alpha. A great feeling of sorrow overcame Mark as he realized that within a few centuries their kind would no longer exist. Then he remembered that unless they became extinct he and the world he was born into could never be. It was one of the many things that bothered him about time travel it forced him to be apart of something that in real time he would be against. It made him feel like an uninvited guest at a party. However, observing history at first hand was of some consolation. When the Neanderthals returned the next day to further examine the Alpha, robots stunned and carry them into the laboratory where doctors altered their DNA. After changing their DNA robots returned them to the place they fell when stunned. As the Neanderthals regained consciousness they slowly rose to their feet and with a confused look they ambled off into the woods with no memory of what had taking place.

CHAPTER 5

Captain Walker and the Planing Committee held a meeting to discuss the issue of the passengers and crew staying with the Alpha or returning to real time. Professor Scott: Captain before we return anyone we had better find out if our generation still exists. Explain Professor? When we exceeded the speed of light it is possible we altered our own Real Time. I believe it would be wise to send someone back to check it out. I agreed Professor. Mark; take one of the scout crafts and check out the Professor's theory.

With his robot crew Mark headed back to the time of his generation. On arriving at where the spaceport had been Mark was surprise to see some odd looking animals grazing. He decided to scout the Earth and see if he could find any humans. When he was through scouting he realized there no humans and could only find 30 of the 25 million species that had inhabited the Earth. The nearest species to human was an ape like creature.

At the Committee's debriefing when Mark revealed the Human Race did not exist at the Real Time of their generation a silence as if the silence before a storm fell over the meeting. Then Professor Scott addressed the meeting. Gentlemen; I'm sure you are now aware that we not only have to see that Neanderthal and Cro-Magnon exist, but we have to monitor human evolution to assure the existence of our generation. Mark. Yes Captain? Mark I'm having two of the three scout crafts manned by robots. I want you to use them to monitor human evolution. You will have full authority to alter evolution to assure the existence of our generation. If you succeed you are to report our situation to the United Nations. In the mean time we'll go back in time and alter the DNA of the southern ape again to try and bring

about the existing of Cro-Magnon. Meet us there when you

return.

CHAPTER 6

On this return trip Mark found life in the 25th century unlike depicted in history. The mode of transportation was dog drawn carts instead of cars. Ox drawn carts instead of motor trucks; and even the smallest village had coal driven railroad train service. One of the oddest things was a huge whale farm off the cost of California. Since Man had not discovered the use of crude oil, gas or electricity, whale oil was the main source of lighting. Antibiotics were unknown and a simple infection could cause death. For this reason the human population was only a tenth of that in recorded history.

Scouting back through the pass Mark discovered the cause for the deviation. In 1750 a Pope by the title Thomas the 6th ordered there be no more scientific research. The

Pope believed such research was against the will of God. R99 I have a job for you. We are going back and pay a visit to the mother of this Pope Thomas 6th before she became impregnated with him. I want you to create a situation whereas the father and the mother never meet. Yes Captain. Shortly R99 returned. Captain I created a problem so the Pope's father never met his mother the Pope never existed. After altered the 1750 evolution of human history, Mark set out on a trek to the 20th century. The 20th century was crucial because it was the century man discovered the use of the atom. On arriving in the year 1946 Mark was surprised to see the German fag flying over Washington DC. Listening in on radio broadcast he discovered that Germany had developed the atom bomb ahead of the United States. They had used the V9 rocket a two-stage version of the V2 rocket to deliver the atomic weapon system to the United

State. After they destroyed New York and Baltimore with its steel mill and shipyard at Sparrows Point the United State sued for peace. The winning World War 2 by Germany has left me with a problem R99 that I don't know how to solve. I have a suggestion Captain. What is it? I could use one of the scout craft to go back in time altar Einstein's childhood. How would that solve the problem? It would cause him to grow up to be the Einstein that writes President Roosevelt. What good would that do? The letter Einstein wrote Roosevelt was the importance factor in the President rushing the research on the atom bomb. OK R99 do it. Mark watched as the Swastika on the flag suddenly change to the Stars and Strips of Old Glory. I see you were successful R99. Yes, Captain but it wasn't as easy as I had first computed. I also had to altered the childhood of a

Rodney C. Terrell

famous German scientist so he would convince Hitler the

unimportance of atom bomb research.

CHAPTER 7

The next important date to check out was 1957 the year man first sent an object in orbit. When Mark arrived in 1957 instead of the Russian's sputnik orbiting the Earth there was a space station. Something is wrong R99; this is the year 1957 and history showed there was no space station at this time! Mark discovered the caused of this diversion from history. It seems there was this American rocket scientist that took the advice of the world famous 1920's aviator to engage engineers to assist him in the design his rocket. By 1938 the United States had a 2-stage rocket that was able to orbit the Earth. With the ability to send rockets too anywhere in the world the United States gave Japan an ultimatum. If they didn't pull their troops out of China and return all conquered territory their home island would be destroy. Hitler fearing the same thing would happen to Germany changed his mind about world domination and World War 2 never took place. A dilemma faced Mark. To alter this version of history million's people would die in World War 2. Mark decided to travel forward in time to see what would develop if he didn't alter this event.

By the year 2000 Man had establish colonies on the Moon and Mars. So far so good he thought; now less see what it's like in 2200. Descending through Earth's cloud's alarm went off all over the scout craft. What is causing the alarms to go off R99? Radiation. Quick get us out of here and take us back to the 1930. Yes Captain. R99 I want you to take one of the scout craft and monitor from 2200 to 2300 to find out what takes place with this version of history. Within a moment the scout craft returned and R99 told Mark what it found out. In 2165 China exploded the first atom bomb. By 2170 the United State had developed it's own atom bomb. The human population in 2185 had reached 30 billion. With such a huge number the Earth was incapable of providing food for them. In 2191 a food war broke out between China and the United State. After they finished firing atomic missiles at each other all life on Earth became extinct. All that remained of the Human Race were the 65 million that made up the colonies of Mars and the Moon.

Mark wondered what effect this version of history would have on his generation. R99 I want you to make a time probe to monitor this history to see what effect it will have on my generation. R99 had no sooner disappeared than it reappeared. Captain I have the information you wanted. O.K. R99 less have it. In the year 2410 the Moon colony was running out of water. Since ever thing on Earth including the water was contaminated with atomic radiation they had no choice but to appeal to the Mars colony for some of their water. When Mars refused to share their water, the Moon colony attacked Mars with missile. However, before their missiles reach their target Mars launched its missile and when the war was over no one of the Moon colony was left alive. The human population now consisted of the 10 thousand on Mars that had survived the war. In 2500 the water ran out on Mars and the Human Race became extinct. Well, R99 I guest that answers the question about altering the rocket scientist reply to the aviator. You know R99, being in charge of mans evolution some times is a depressing job. First, it was the Neanderthals now the though of all thought millions dying in World War 2 may him even sadder. However, it is the only way to assure the Human Race would continue. This presents Mark with an unusual problem since history had shown the rocket scientist and aviators actual did meet. R99 I want you to go back and alter the rocket scientist's childhood so he becomes the same person as in our history. This way we know he won't consider the famous aviator suggestion.

When Mark arrived in the 24th century he found the Earth swarming with people. Although some countries had laws governing its population growth other did not and when the Earth's population grew to fifty billion food wars became a common event. Earth's over population led to the cutting down of the forest to plant food crops. With destruction of forest that recycles carbon dioxide into oxygen the air became unhealthy. Once it was realized more people were dying from the bad air than starving to death there was only one thing left to do, build dome cities and farms. Billions of people either starved to death or died from the bad air before the last domed city and farm had been built.

CHAPTER 8

It was now time for Mark to find out if his generation existed; so he set the time on the computer guide system to arrive moments after the Time Craft Alpha had left. As the scout crafts gently lowered its self onto the same New York space port pad the Alpha had been launched Mark was surprise to see a huge crowd of people looking up at the scout craft. Then it dawns on him it was the same crowd that was there to see the Alpha lift-off only moments ago. Entering a pressurized underground corridor he stepped onto a people mover that carried him to the space port main terminal where he found himself surrounded by TV reporters that realized the scout crafts were from the Alpha. He assured them the passengers and crew of the Alpha was safe and well. Then he commandeered the limousine used by the Planning Committee only minutes before to transport them to the spaceport. After instructing the driver to take him to the United Nation building he used the cars VSC to call UN President O'Connor and notified him of his arrival. Arriving at the U.N several security guards escorted him through the crowd of reporters. President O'Connor? Yes. I'm Mark of the Alpha. Come lets us go into my office. As he entered the vast circular U.N conference room Mark was impressed by the huge globe of the rotating Earth located in its center. Around the Earth's globe orbited the Moon and Mars with little domes on them representing their colonies. On reaching the office President O'Connor pointing to a chair "please be seated". I must say I'm a bit confused; you said Project Escape was successful and yet you returned only moments after your departure. I would like to know how this is possible? You must understand that time travel gives one the ability to subtract or add to Real Time. So in order for me to return to this moment I simply subtract the time since the Alpha left the launch pad. I then put the information plus the time it takes the scout craft to arrive at this moment into the scout craft's computer guidance system. So; what you are saying is you could arrive before you left. I could but I would have the same problem materializing as those on the Alpha back in time. Materializing! Yes; You see a time traveler can not physically exist once the person reaches their destination unless they existed at that point in Real Time. Also you cannot physically be in two places at the same time. I'll give you an example; although I could go back to the time that I was being born an observer my birth I would not be able to physically observer the birth. The reason I am able to be setting here talking to you is that I synchronize the scout craft's arrival at the point the Alpha first broke the speed of light barrier thus reversing the physical altering factor. If a person cannot physically exist once that person travels back in time then how can you say Project Escape was successful? Although we are unable to physically exist when we reach our destination you must remember since we exist outside of Real Time we have an eternity to solve any problem. How will that help those of us who exist in the present; we only have a 25 more years before our world cease to exist? You remember what I say about those of us that live outside of Real Time? Yes. Then you must realize in Our Time we will be able solve the problem and then return to this time and transport every one back in time. This time traveling is a bit confusing. Tell you what; I'm going to call the U.N Security Counsel into session tomorrow and I would like for you speak to them.

Meanwhile, I'm assigning you a couple of guards and you can have the use of my helicopter to avoid the crowd out front.

Climbing aboard the helicopter Mark instructed the pilot to take him to the Dome Hotel on Fifth Ave. As the helicopter settled down on the hotel rooftop's helicopter pad Mark had one of the guards register for a suite of rooms in his own name. After all if the news reporters discovered where he was staying there would be little chance of him getting any rest.

As soon as Mark settled into the suite he called the two top scientists in their field, Houser of physics and Morgan of biology. He explained to them the problem Professor Andrea and his staff faced with finding a way to materialize and ask them to join the professor in his research. Realizing that Professor Andrea would need all the help he could get to find a way for humans too physically exist outside the Alpha Morgan agreed to join the professor. When Mark told Houser that Morgan was joining Professor Andrea's team Houser also agreed to join. With Houser and Morgan agreeing to return with him Mark decided to call it a day. He was about to call room service to have them send up some food when one of the guards entered the room and informed him there was a woman reporter outside that wanted to see him. Is she alone? Yes. Well, I suppose I should see her if I don't she could return with a hold army of reporters. As he got up from the chair to greet her he became transfixed; she was the most beautiful woman he ever seen. Just looking at her made him feel good all over. As he looked into her eyes he sensed that he also met with her approval. He wanted to tell her how beautiful she was but decided against it for if he was any judge of character she'd probably been bored to tears by so many such comments. He didn't realize he had been standing there gawking like an infatuated schoolboy until the guard asked if he wanted to be alone. Yes, I thank we can trust; by the way what is your name. Gladys Shorter; I'm from The World newspaper. What can I do for you Mrs. Shorter? Or is it Ms? It's Ms. How did you find me? I heard what you said at the spaceport and it seemed to me that you were not telling the hold story. I decided to follow you. On seeing the helicopter leave the U.N building I figured you were giving the crowd the slip. I called my newspaper editor and had him send someone up to the roof of the newspaper building with binoculars and to let me know where a helicopter with U.N letters landed. How did you know which room I was in? I figured you would send one the guards to registered in his own name. When the desk clerk turned his back I scanned the register with my VCS and sent it to my newspaper computer where the names of all the U.N. guards are store. You are a very smart woman Ms Shorter. Why thank you. I was just about to order some food would you care to dine with me? It's the lease I can do for all the trouble you went to find me. Dose that mean I don't get the story? I'm afraid so; as much as I would like to I can't give you the story; I can't until after my U.N. speech tomorrow. Well, since you won't give me the story dinner as a consolation is better than nothing. The disappointing look of on her beautiful face was too much for him. I tell you what; if I give you the story will you promise not to print it until after my speech tomorrow? I promise. Shall we shake hands to seal the deal? As his hand touched hers a thrill surged through his body and for the first time in his life he was consumed with the desire for a woman. Looking into her eyes he could tell that she also desired him. Then his lips met hers and they were over come with passion. At that moment Mark knew he wanted her to be a part of his life.

During dinner he became more of the interviewer than the interviewed. How long have you been a reporter? Three years. You look so young for the job. I was able to get the job because my foster father worked for the newspaper. Your foster father. Yes, A meteorite killed mother and father when it struck their room while they were vacationing at the G1 space hotel. He could see she was about to cry so he decided to change the subject. Pouring champagne in their glasses he suggested having their drinks on the balcony. As they gazing up at the sky the Moon slowly appeared from behind the main dome support building in the center of the city giving off a golden glow due to the polluted air on the outside. Back at the Alpha the

Moon looks much different. Tell me about that world. As he related to her the sights and sounds, he could see that her curiosity was still not satisfied. Would you like to go back with me and experience it for yourself? O yes! Well, why don't I talk to your editor after all I'm sure he can see value of having a reporter give a first hand report.

After informing the U.N Security Counsel of the status of Project Escape Mark met with Gladys's newspaper boss Mr. Canon. Canon thought sending a reporter to get a first hand report on Project Escape was a great idea. However, he felt a reporter with more experience than Gladys should get the assignment. When Mark told him how smart Gladys was in tracking him down at the hotel and that he would consider it a personal favor to him. Canon agreed to let her do the story. When they were in the elevator after leaving Canon's office Gladys turned to Mark and gave him a hug and kiss. Thanks for going to bat for me. Your kiss is amply reward. Mark noticed a huge crowd of people waiting to board the scout craft when Gladys and he arrived at the spaceport. Far more than the scout craft could accommodate on the trip back to the Time Colony. Mark's fear of a riot passed when he was able to convince the crowd that another craft would arrive shortly. He didn't tell them it would be the same craft.

CHAPTER 9

Back at the Time Colony Mark reported his finding to Captain Walker. Afterwards he spent the next two weeks relaxing. Later he helped Gladys with editing and copying the videos for her newspaper story. When Gladys was finished putting the story together she decided to spend several months at the Time Colony before going back to her Real Time. After all she could always go back to within seconds of when she left regardless of how long she spent back in time. During that time she and Mark were inseparable and although he was 43 years old and she only 18 he decided to asked her to marry him. She agreed to marry him after she returned from delivering her story to the newspaper. Mark gave Gladys a hug and kiss as she joined the others on their trip back to Real Time. I'm going to worried until you get back. There no need to worried I'll be back in a moment. Mark had only been standing in the disembarking area a couple of minute when he saw the scout craft returning.

As Gladys disembarked a puzzled look came over his face; she looked much older. Suddenly it dawned on him while was standing there she must have spent years in the future. He was no longer looking at a 18-year-old girl but a woman of 30. What happen? It was terrible when I returned home ever thing had changed primitive people populated the Earth. There were no dome cities or farms instead the Earth looked much like now. Some of the people aboard the scout craft decided they wanted to settle there and live out their lives. When others and I try to convince them that the Earth was still doomed we were imprisoned. It was only with the help of a sympathetic guard that we were able to escape and gain control of the scout craft. Now that I am older due you still want to marry me? Of course; you are still the love of my life at the debriefing on her trip she repeated her story. President Walker: This is terrible it means we have to alter evolution on every trip back to Real Time.

One day President Walker informed Mark he wanted to see him. Congratulation on being named President. Thank you Mark but its only temporary until the election. Now as to why I sent for you. Mark while the scientists are trying to find a simpler way to return to our generation the planning committee and I decided we should have someone explore time. I can think of no one more fit for the job than you. I'm asking you to be Captain of the Omega; a new Time Craft that I am having robots build. If accepted you will be reporting directly to me. Because of the great distance on some trips it will be necessary for you to be placed in a state of suspended animation. So there you have it; what do you say? Can I let you know after I've discussed it with Gladys? Sure call me tomorrow. That evening at dinner Mark told Gladys of President Walkers idea. Well, what do you think? I think it's great idea proving I can go with you. As Captain of the Omega I wouldn't have it any other way.

CHAPTER 10

For your first trip I want you to go back in time to when dinosaurs roamed the Earth and find out why they became extinct. On arriving Mark placed the Omega in hovering mode a mile above the Earth. As Gladys and he were looking down at the Earth R99 brought to their attention that a huge asteroid was heading directly toward Earth. I think it would be wise to witness the event from a safer distant out in space Captain. I agree with you R99.

A great cloud of dust was thrown up into the atmosphere as the asteroid struck the Earth, blocked out the view of the surface. R99 take use five hundred years into the future. On arriving it surprised Mark to find that some of the dinosaurs survived the asteroid strike. Now he had to find out what did cause their extinction. R99 take us to the future. It was several stops before they discovered a dead T. Rex that didn't show any sign of a wound. Mark sent a robot to obtain sample of the corpus to determine what caused its death. It didn't take long for the robot to determine the dinosaur died from a virus. Mark decided to travel further into the future to find out what effect the virus had on the dinosaur population. Before landing the Omega Mark ordered R99 to send out the scout crafts to search for dinosaurs. Although there were many smaller species and some birds the only sign of dinosaurs ever existing was their bones. Mark found it hard to believe that a virus could wipe out the dinosaurs and yet have no effect many of the other species. He decides to have R99 round up some of the species that survived and compare their immune system with that of the dinosaurs. It soon became apparent that their immune system was stronger than the dinosaurs thus enabling them to fight off the virus. After R99 and the other robots finished cataloging the difference life forms Mark headed back to Time Colony.

When President Walker heard it was a virus that killed off the dinosaurs there was a surprised look on his face. Imagine that; the largest form of life that every roamed the Earth killed off by a virus one of the smallest form of life. Well, that blow's a hole in the theory that an asteroid was the cause of their demise. However, since it took place about the same time it's an understandable mistake.

CHAPTER 11

Mark your next time trip will be to observe the solar systems creation. Since you will be traveling a great distance back in time it will be necessary for you to spent most of the time in a state of suspended animation. Who will control the Omega? Robots; they will not only control the Omega but also repair and make improvements on it and themselves during the trip. Standing by the duel life suspension capsule Mark and Gladys hugged and kissed and then climbed into the capsule.

When Mark awoke he found himself staring up at a creature except for its size look nothing like the R99 what seamed only a moment ago had sealed them in the suspension capsule. Who are you? I am ROD head of all robots. Who named you ROD? I gave myself the name I took it from the words (Robot Operational Director). How do you feel Captain? A little dizzy. I'll have a robot bring you and the lady something for it. Looking at the time read out on the life suspension capsule Mark turned to Gladys and putting his arms around her and gave her a hug and kiss. That's to show you millions of years has not changed my love for you. O! Mark I was so afraid I would never see you again. Its lucky we have a robot crew, I hate to think of how many mistakes a human crew would have made in that time.

Switching on the Omegas view screen Mark and Gladys saw not one but twin suns with no moons or planets revolving around them. ROD can you explain this? No Captain. Then I want you to send out a robot time probe to determine when the solar system was first formed. The probe having an antigravity drive returned only a moment after its launch. It reported the solar system came into existing when the smaller of the two suns exploded.

On arriving at the solar systems creation Mark and Gladys witness a sun flare of incredible size. ROD can you tell me what's causing that huge sun flare? It is due to the matter from the exploded smaller sun falling back into the sun and then being then flung back into space by the sun's rotation. ROD can you tell me why the planets are made up of different elements? The outer planets consist of light elements because they were able to escape the sun's gravity with less effort than the heavy element inter-planets. Although they consisted of the same heavy element as the inter-planets the asteroids being so far from the sun, were ejected from it in little blobs instead of the larger ones that made up the inter-planets and moons. The shape and

Rodney C. Terrell

smooth surface of the asteroids are due to the fact they exist

in the same form as when first created.

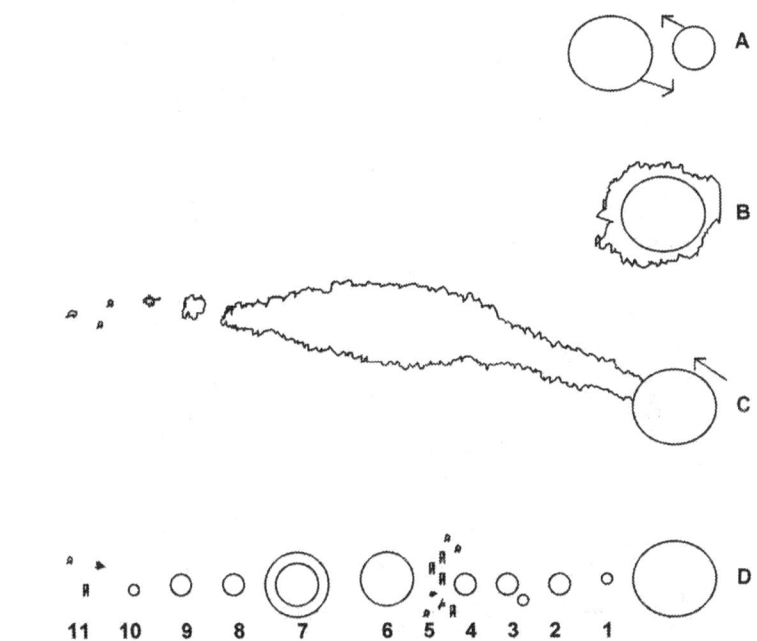

1.Mercury. 2.Venus 3.Earth 4.Mars 5.Asteroids belt 6. Jupiter 7.Saturn 8.Uranus
9.Neptune 10.Pluto 11.Comets.

A. Our sun and it's smaller twin revolving around one another .

B. The gravitational pull of our sun overcome the orbit of it,s twin causing
   it to fall into our sun and explode.

C. Because our sun was spinning at a higher speed than now the matter from its twin
   is flung out into space by a incredible large sun flare from which the planets, moons
   asreroids and comets were created.

D.Since the lighter matter of the sun flare easily escaped the sun's gravity it
   became the outer planets. The heavy matter being harder to excape the sun's
   gravity became the inter- planets. The asteroid belt that lay between the light
   elment outer planets and the heavy elment inter-planets was unable to form .
   into a planet because being of heavy element and so far from the sun was flung
   out from the sun in little blobs instead of large ones that made up the inter-planets.
   When you look at the smooth surface of a asteroid you see the same shape
   it had when it and the solar system was first form.

CHAPTER 12

Just think; Our sun's twin being responsible for the solar system existence exclaimed President Walker. You know Mark, it makes sense when one realize 75% of sun are multiply suns it is understandable that single ones at one time would have a companion. On learning how ROD had evolved. Mark if ROD is as intelligent as you said, maybe it knows a way we can physically exist outside the Time Craft, have it come over here and less find out. ROD come over here President Walker wants to ask you something. ROD do you know if there is a way members of the Time Colony can physically exist in this time? It is not possible for me at this time to under stand how to solve that problem Mr. President. Tell me ROD do you know a way that we can return to our real time without altering evolution? No, you see there is an infinite number of possible futures this is because when someone is born or dies our future becomes altered. This means that there is only one future that leads to your generation. However, because of the improvement I've made on myself during the last time trip I can make it a lot simpler to alter evolution

For the second time Gladys was boarding a scout craft head back to the future to deliver her newspaper story. Why not come with me Mark? I wish I could but President Walker has ordered me not to because if something happened to me it would jeopardize his time exploration project. Why don't you just send in the story? No I owe it to the editor to take it back in person. Besides with ROD on board every thing will be O.K. All right I will be waiting here for you in the docking bay's bar. He had just begun to sip a Scotch and soda when the bartender came over to the end of the bar where Mark was seated all excited. Did you hear that? Hear what? They just announced the arrival of other Time Craft's; about that time the bars VCS began to beep so the bartender went and answered it. Is your name Mark? Yes. It's for you. Mark? Gladys where are you? I'm aboard one of the Time Craft that just landed. Are you all right? Yes. I'm about to board a scout craft and should be there shortly.

How come all those Time Craft? After I delivered my story to the newspaper I went to see my parents. After several weeks I convinced them it would be better to come back in time than spend their life on a space ship hoping to find a planet safe from the Black Hole to settle on. I had just convinced them and was getting ready to return when it came over the news that latest calculation showed it would be impossible for any more space ship to escape the Black

Hole's gravity with the ion drive. When the people received the news they all headed for the scout craft and it was destroyed by the mob fighting to get aboard. For 5 years Robots worked round the clock before all the spaceships were converted to Time Craft's. To avoid mobs fighting to get aboard a Time Craft the decision was made not to launch any Time Craft until all of the spaceships could be converted. It makes me feel ashamed to have been setting here at the bar having a drink while you were going through all that turmoil.

CHAPTER 13

Of all the spacecraft's converted to Time Craft's only three hundred elected to travel back in time and join the Time Colony. On landing President Walker invited the Captains to meet with him on the Alpha to setup a government. They agreed that Walker should continue on as President until there could be a general election. For the safety of the Time Colony the three hundred Time Craft's were to consist of groups of ten placed in a circle at difference locations around the world. The Time Colony now consisted of thirty communities of a 10,000 each. The Time Craft being three miles across with ten levels not counting the observation bubble on the top made ample room 100,000 people. Looking down from above the community looked like a giant wheel made up of smaller wheels. Counting the 10,000 aboard the Alpha the Time Colony now had a population of 3,010,000.

Although the Omega was only a mile and half from community #1 where President Walker and the Planning Committee ran the government it was still necessary to use one of the scout craft to get there. Mark I know you haven't had much time to enjoy married life so take off a couple of months. In the mean time I want ROD to reproduce a robot with its abilities. By the way what is the next assignment Mr. President? I want you to find out if the Big Bang wrought about the created of the Universe.

While ROD was working on the Presidents project Mark and Gladys used one of the Omega's scout craft to visit her parents. Later they visited the difference settlements. The last settlement visited was in North Africa in what would in the future become the Sahara Desert. By the time they arrived at the scout craft-docking bay the settlers had received the news they would be prisoners of the Time Craft. Are you Mark? Yes. I'm Ryan head of security you got to get out of here before the people find out who you are. Why? Some of the people realizing they can not physically exist outside the Time Craft are in bad mood. As their scout craft arrived at the Omega's docking bay Mark saw ROD making its way toward the Omega from community #1. ROD how come you left the community? The mob has taking over and forced the President to agreed to join the others searching for a planet safe from the Black Hole's gravity. The President instructed me to tell you he wants you to complete the Big Bang project. I think we should leave before the mob attempt to take control of the Omega. What about my mother and father? I would advise we leave right away Captain. I think you are right ROD.

Gladys you can contact your parents once we are away from the area. As the Omega began to take off there was a flash. Captain we are being fired at should I have the staff retaliate? No, ROD the force field shield will protect us. When the Omega was a thousand miles above the Earth Gladys contacted her parents. Mother are you and father all right. Yes, Are you and Mark going to be joining us? Maybe later but first we are going on a time trip for President Walker. You and Mark take care of yourself we long to see you again.

CHAPTER 14

Even at the tremendous speed of the Omega it was estimated to take 500 thousand years to travel back to the beginning of the Big Bang. Although it was the second time Mark and Gladys had used the life suspension capsule they felt uneasy because of the time they would be in a suspended state. Compared to this time trip their first trip was like a short nap. Laying side by side they gave each other one last look and then as ROD sealed the capsule they drifted off to a long night sleep.

When Mark awoke from his suspended state he looked up from the life suspension capsule and saw a creature except for having gray metallic skin instead of white it looked like his twin. Don't be concern it's me ROD. I detect your system is not up to par Captain. Yes I do feel a little dizzy. I'll get you something for that. Within a minute

a robot walked in with two cups handing one to Mark and

the other to Gladys. Drink this it will clear your heads. The

drink taste different from one you gave me the last time.

What is in it? Micro medical robots. How do they help?

They repair any cell damage or chemical imbalance in your

body that you may have incurred during your suspended

state. How long will they stay in our bodies? For as long as

you wish to live. What! What do you mean as long as we

wish to live? Yes, Captain your bodies won't get any older

than they are now. The Micro medical robots will

regenerate your cells, organs, destroy bacteria, virus, heal

wounds, etc. How did you get the other robot to bring you

the drinks without using your VCS? I have created a way to

communicate telepathically with the other robots. A minute

later another robot appeared with two pea size golden orbs

handing one to Mark and the other one to Gladys. By wearing the orb in your ear you will be able to communicate telepathically. By simply closing your eyes you will be able to see whatever I am seeing at the time. What else have you learned? Much more than your brain is cable of computing. Why did you model yourself after me ROD? I figured that since I was doing your job I should up grade my body to look like yours. It's lucky for you I didn't have any hair; that would have created a problem for you. O.K; but why have you changed the Omega's shape and the time read out shows we've only been asleep 89 thousand years you was not supposed to wake us before we reached our destination? First: We have arrived at the beginning of the Big Bang. Second: It was necessary to change the Omegas hull using a super metal and shape to a to withstand the extra speed from up grading of the it's antigravity drive. How did you do that? About 30 thousand years ago I evolved myself to the point where I'd gained the ability to alter matter by using thought. From now on ROD when I am in a suspended state I want you to revive me before you take any action other than that pertaining to navigation. Yes Captain.

Mark and Gladys could not believe their eyes although they were ten light years from the Big Bang; the brightness was like that of a atom bomb explosion. ROD on the trip have you evolved to the point you know what caused the Big Bang? Yes, it's the results of what happens when the first Galaxies created by a previous one escaped the blast force of that Big Bang. Once they escaped the Big Bang's force their gravity pulled them together causing them to collide. As the stars of these colliding Galaxies burn out they created a Super Black Hole that pulls in most of the other Galaxies. Once most of the matter of the Universe has been compress to a critical point by the gravitational force of the Super Black Hole there is another Big Bang and the process repeats its self. How did the first Universe come into existence? There is no beginning or end Captain. Are you sure! Yes, since the Universe is made-up of matter and energy the only beginning or end is when one is converted into the other.

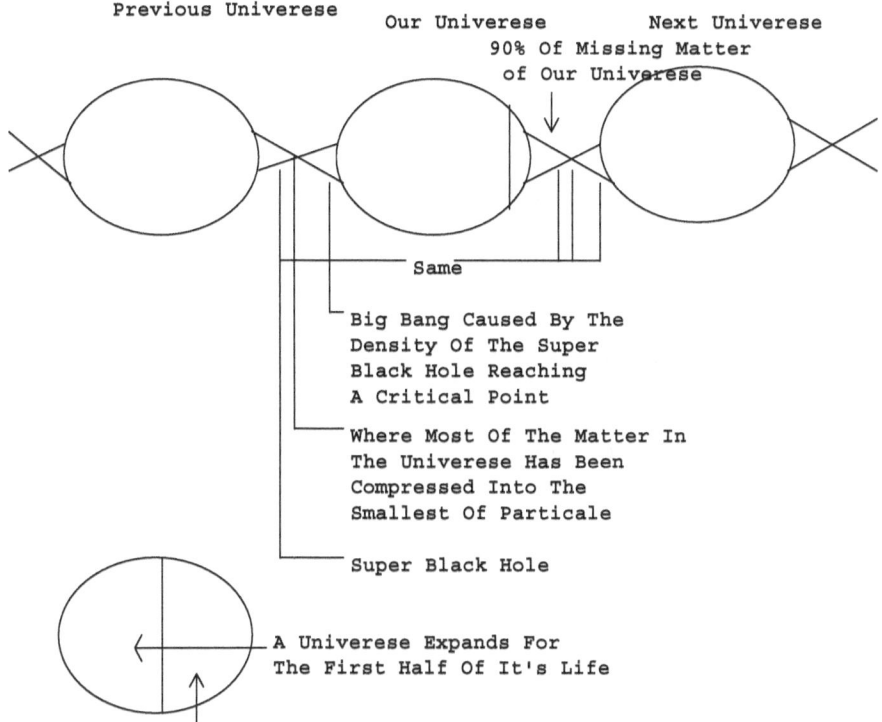

Previous Univerese

Our Univerese
90% Of Missing Matter
of Our Univerese

Next Univerese

Same

Big Bang Caused By The
Density Of The Super
Black Hole Reaching
A Critical Point

Where Most Of The Matter In
The Univerese Has Been
Compressed Into The
Smallest Of Particale

Super Black Hole

A Univerese Expands For
The First Half Of It's Life

As The First Born GaLaxies Overcome The Force Of the Big Bang And
Are Drawn To One Another By Their Gravitational Force Become
A Colliding Galaxy As The Suns Of The Colliding Galaxy Die
They Create A Super Black Hole That Eventually Swallow Up Most
Of The Other Galaxsies In The Univeres.

The Lifetime Of A Univerese Can Be Determine By Computering
The Amount Of Missing Matter It Should Contain
Example:
If 50% Of Matter Is Missing From It And The Big Bang Occured
15 Billion Years Ago; Then The Time Between Each Big Bang
Would Be 30 Billion Years. Since The Big Bang Occured Before
All The Galaxies Fom The Previous Universe were Drawn Into The
Super Black Hole That Brought About The Creation of Our Universe
Those Galaxiese will be Older Than Our Univerese

On our way back Gladys would you like to stop off and observe the beginning of life on Earth. Yes I'd like that. ROD I want you to change the Omega from its tear drop shape back into the original saucer shape. Why Captain? So we will be more recognizable in the event we come across a Time Craft from the Colony.

CHAPTER 15

The Omega arrived about the time when it was assumed life first appeared in Earth's oceans only there was no life. ROD take a scout craft and search through the future until you find out just when life did begin on Earth. When ROD returned it informed Mark except for vegetation the Earth was barren of life. How far did you travel into the future? All the way to the time your generation existed in history. Rod; I don't understand. I am sorry Captain but it seems we have returned to a different Earth than the one we left. Does this mean Gladys and I will never be able to join the others? There is a way. How? On the trip back I evolved to where I can create primitive life forms. You never cease to amaze. You know what they say Captain in time all things are possible. ROD you are becoming increasingly human. I will never be human Captain because I do not have emotion nor any desire to have any. Without emotion ROD you can never know love. I serve the purpose in which I was created

ROD; explain to me how you can create life? First, I wait till the Earth has cool to the point where the heat won't destroy the DNA that I will create. Once life appears I monitor it to assure it evolves into the desired life form. What if it doesn't? Then I will alter the DNA. Since you have evolved to such a high degree I wonder if you know an easier way to fine-tune mans future to insure the existence of our generation. Yes Captain; with your permission I can improve the orb you wear in your ear to make it possible for you to see and know what a person is seeing or thinking at any time in the past or future. Also when they are asleep you can through their dreams alter or cause others to alter the actions of anyone or anything that would attempt to change Mankind's destine. Want the people know they are being monitored? No, they will have an awareness and many beliefs to explain it however they will not know. What will happen to all the men and women that are alive when we go back in time and alter their time to coincide with our history? They will still be born and live out their lives. Won't they remember having lived this life? Only when the event they have lived coincides with that event in real time history it's referred to as deja vu. O.K. go ahead and make the improvement on the orb.

Mark if something happened to ROD how would you create another orb? I see what you mean Gladys. ROD I want you to up grade two of the R50 to your ability. With Mark new powers it would no longer be necessary for him to have ROD genetically alter the subject. Now through dreams and thoughts he would be able to alter the action of those that would cause a deviation from the events that historically occurred.

ROD is there more you can tell me about dreams that I should know? Yes, think of the subconscious mind as a TV station able of transmitting and receiving throughout time. At night when a person is asleep he or she is receiving on the subconscious level thoughts and ideas of persons awake that's living some where in the past or future. It's this ability of humans that makes them the only species able to created thing that never existed before in their time. Since the subconscious mind doesn't have a turner like the TV receiver it picks up signals randomly. This is why one minute you can be dreaming of one thing and then suddenly you find yourself dreaming of something else. In the case of precognitive dreams it's the persons future brain sending a picture of an event that is taking place back to their present brain. On rare occasions this can take place while the person is awake. It's this ability of a human's brain that made it possible for your generation to find a way to time travel. Other species rely only on their instinct and are destiny to become extinct unless humans save them. Whereas the human subconscious mind being cable of receiving new ides from other human beings through out time allows the Human Race to alter its future to avoid extinction.

To Mark's Generation

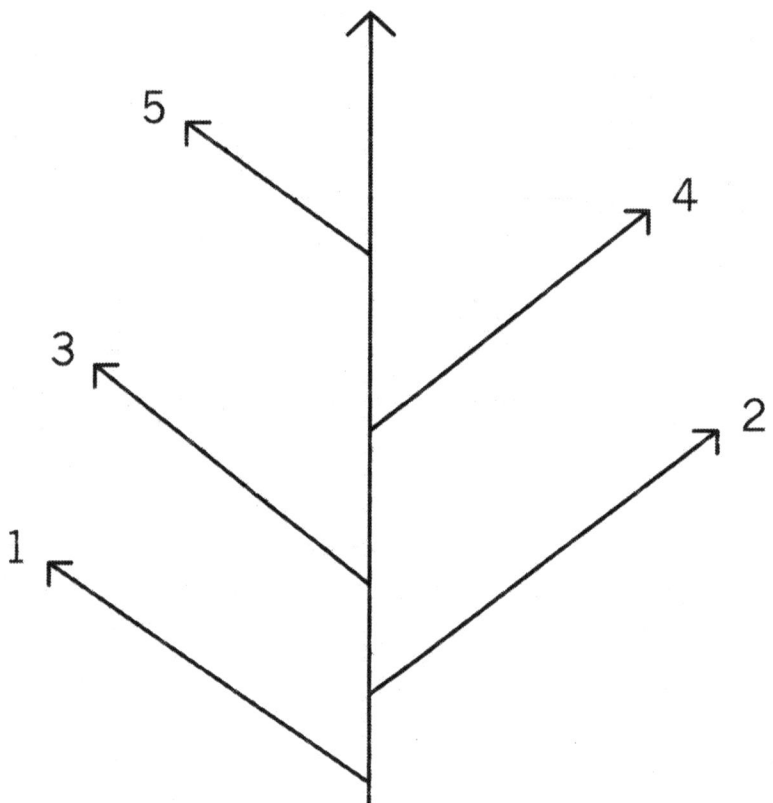

Five of an infinite number of possible
future's that bring about
Human extinction

ROD if what you said about deja vu is true how do you explain the deja vu time travelers experience? It means they are reliving a moment that they lived in one of their pervious lives. Explain. Say you have a billiard table with on one side ten red billiard balls numbered one to ten and on the other side of the table ten white billiard ball's number eleven to twenty and mix them up. Now if I was to pick up the table and shake it and set it down enough times eventual all the balls will wind up in the same position as they were in the beginning. The same laws of physics apply to the Universe. As the Universe becomes renewed an infinite number of times by way of Super Black Hole's you could pick any moment in time and that moment will occur again. This means that your deja vu experience is from the same event that you experienced in one of your previous lives. You mean I've lived before? Yes, Captain many times before however not necessary in the same Universe and not exactly in the same way in each life. Being a robot I can understand how difficult it is for you to comprehend such a span of time. I'll explain it this way; I estimate it takes 30 billion years for the Universe to complete a cycle say each cycle represents a Z-hour I compute within 3 millions Z-years every atom in a future Universe will be as they are at this moment. Before that happens there will be many variations and some of the events will repeat themselves

ROD I am leaving the problem of creating the different species up to you. Gladys and I are going to start monitoring the Human Race at the beginning of its written history we'll meet you at 06/29/1828 at 4:00PM over central north America. Yes Captain.

CHAPTER 16

In the 3rd century a religious group called the Essences became the worlds major religion. Since it would significantly change history it was necessary to go back to the 1st century and alter the thoughts and ideas of some key people. As the Omega appeared over the Mediterranean sea Mark and Gladys saw for the first time saw an eclipse of the sun that was not through the city dome of their Real Time. After altering the thoughts of the Egyptian Roman ruler and some Essences Mark and Gladys headed for the meeting with ROD. Hovering over the planed meeting point Mark noticed far below some American Indians staring up at the Omega. It reminded him of one of the story his mother told him when he was a child. The story was about the North America Indians and their legend of how a huge round object came down out of the sun. Suddenly Mark realized the Omega must be the object of that legend. Another legend came to mind; this one occurred in biblical times when a member of an ancient religion described an object in the sky that looked like a wheel within a wheel. Looking up at round shape of Omega and it's scout craft, in their docking bay beneath to someone of that time it would look like a wheel within a wheel.

They had been waiting only few minutes when a scout craft suddenly appeared above the Omega. At first Mark thought it was ROD arriving, until he realized the marking on it wasn't of the Omega. He decided to try and make contact. This is Mark of the Time Craft Omega. The face of President Walker appeared on Marks VCS. President Walker! How did you find us? It wasn't easy. Ask him how are my parents. Gladys want's me to ask you how are her parents? I hope they are all right. What do you mean? After the rebels took control of the colony they set out to join up with those that used the ion drive spaceships. When they located them they found them loosing the battle to escape the Black Hole's gravity. Once I made them understand it would be necessary to travel faster than the speed of light to escape the Black Hole they agreed to let the robots convert their ion drive to antigravity drive. When the rebels realized the whole Universe of that time was going to be destroy and what they were trying to accomplish was impossible they restored me as President. The Planning Committee and I voted that we all should return to the Time Colony settlement.

Every thing was find until the R99B robot that ROD upgraded malfunctioned and destroyed all the other robots. Since the R99B was so superior to our scientist there was no way to bring it under control. I figured that with the time ROD had to evolve during your trip to check out the Big Bang it would be able to control the R99B. I sent ROD on a mission and it's due to meet us here any minute. As Mark was speaking ROD,s scout craft appeared. Mark; have Rod's scout craft hover long enough for me to use the empty docking bay to come aboard. Yes sir. After President Walker had disembark and his scout craft disengaged from the Omega and headed back to the Time Colony ROD's scout craft docked and the Omega also set course for the Time Colony. On the way Mark and the President briefed ROD on its creation project.

I was unsuccessful on my first try in causing the species to evolve into humans. This was due to the Earth being so thick with vegetation. I solved the problem by creating dinosaurs to thin out the vegetation. I was curious what the dinosaurs would evolve into if left alone. To my surprise many of the smaller ones evolved into birds. Although the asteroid killed off some dinosaurs they survived for 6 million more years until they had destroyed all of Earth's vegetation. With no more vegetation the dinosaurs and all other land species died off. I then went back in time and created the virus that destroyed the dinosaurs. After the virus destroyed them I took a trip into the future to see the result. It was not what I'd expected; insects had destroyed all other forms of life. It didn't take to much computation to realize the best way to control insects was with birds. Again I went back into the past and create a virus in such a way that it would not be harmful to the dinosaurs destine to become birds. After ROD finished relating its adventure President Walker with a stunned look on his face reached for a chair and sat down. Mark, when I told you to do what was necessary to guarantee our generation would exist I never dreamed you would have to go to such extent.

CHAPTER 17

Arriving at the Time Colony Mark joined President Walker on the Alpha that was being used as seat the of government. With all the converted space ships to Time Craft's the communities now numbered over 5 hundred. While Mark sent ROD to find out why the R99B had malfunction and to neutralizing it, Gladys went to see her parents. Captain I've completed the mission. What caused it to malfunction? It was due to a micro meteor that penetrating its circuit. I had the same problem with one of the robot staff on our trip back to check out the Big Bang. Why didn't you tell me about it? I saw no need to worry you Captain since I had solved the problem. How did you do that? I created a metal so dense that even a micro meteor was unable to penetrate and used it as the outer cover for the robot staff and myself. Then use the metal to create the new robots the R99B destroyed

Mark when you get the time I want you to meet me here at my parents; they living in community #41. I have good news and I want us all together before I tell what it is. Well, it looks like ROD has ever thing under control; so I'll grab a scout craft and be right there. On the way Mark thought after all the bad news he'd had in his time trips he was ready for some good news. what good news? I'm going to have a baby. Sweetheart when did this happen? Remember in the 3rd century when we took some time off our journey to relax. O'yea; you know Gladys we need to take more time off to relax. Mark do you think our child will be able live out side of the Time Craft? Theoretical it should be possible; I'm sure ROD would know. Yes, Captain a child will be able to live outside the Time Craft at the time it is born. Mark; this leaves us with a problem. What problem

Gladys? The environment outside at this time is to hostile.

Don't you think I should have the baby at a time where the

outside environment is less hostile in case it would want to

live outside? Yes, I do but we will have to get the O.K.

from the President. I'm going to take this idea to him and

I'll meet you back at the Omega when you through visiting

with your parents. Gladys how did Mark contact ROD

without a VCS? Taking the orb out of her ear Gladys

showed it to her mother. This allowed me to communicate

telepathically. That amazing do you thank your father and I

will ever be able to have one? In time ever one will have

them.

President Walker I need to see you about a problem.

What is it? I just learned from ROD that a child will be able

to physically exist outside the Time Craft in the time in

which it is born. So how is that a problem with you? Gladys is going to have a baby. Congratulation but you still haven't said what's the problem. The problem is the environment is too hostile at this time. Gladys and I would like to go further into the future before the baby is born. I see what you mean, and if it was up to me I'd say O.K. but this also concerns the Planning Committee so I'll have to get their approval. I am meeting with them tomorrow so I'll inform them of the problem and let you know their decision.

While Mark was waiting for the decision he checked with ROD to see how it and the other Omega robot were progressing with rebuilding the thousands of robots the R99B destroyed. I should be through in 3 days Captain. OK; when you are finish return to the Omega. Mark; President Walker here I got the Committee to agrees that all women who are pregnant have the choice of travel forward in time and sat up a settlement. Thank you Mr. President. Before you and the other leave on your trip the Committee and I voted that ROD should upgrade the R99B to its capabilities.

Mark I just hear on the news they are letting the pregnant women and their husbands that want to set up a settlement in the future. Yes, I know the President informed me. When will we be going? As soon as ROD and it's staff finish rebuilding the destroyed robots. How far into the future should we go? The President and Planning Committee agreed we should use that location of our first attempted Time Colony. You mean Egypt? Yes. The Omega and one other Time Craft was used to transport the 10 thousand elected to join Mark and Gladys.

CHAPTER 18

Arrival at the site Gladys found not the desert as depicted in ancient pictures but a lush green paradise. Mark had ROD and it's staff of robots set about building permanent dwelling for the children for when they were old enough to leave the Time Craft. He and Gladys's son whom they named Adam would be among the first to leave a Time Craft since the beginning of Project Escape. After ROD and it's staff finish building the dwellings Mark had it build a huge Sphinx so if someone got lost they could use it to find their way back to the settlement.

When Adam was 5 years old with ROD at his side he took his first walk out side the Omega. O'mother it is so wonderful out side the air has so many different fragrances and it is so much better being able to touch the little creatures than just looking out at them from inside the Omega. Who was the little girl you were playing with? That was Eve; her father is in charge of the robots cataloging the different species. As the years went by Adam and Eve became inseparable and when Adam was 18 he asked Eve to marry him. Eve agreed to marry him if he would live with her outside the Time Craft. First I want to discuss it with mother and father.

If you and Eve choose to live out side the Time Craft neither you or your children will not be given the Youth Drank that ROD gave me and your mother. Why not father? Because it would alter history and make it impossible for your mother and I to exist and if we never lived then you would not be born. You mean I will die? Yes, it's the price you would have to pay be able to experience the physical sensations and adventures you don't have living in the Time

Craft. What happens when I die? You'll be reborn into another life. Will Eve be reborn with me? In some of your lives she will. Will we look the same in our other lives? Not necessary. How will I recognize her? You will not recognize consciously; however when you look at a woman and get the feeling that you have know her before she could be Eve. We have a saying for it; we call finding our soul mate. I had that feeling when I first saw your mother.

What did you decide Adam? O.K. if being a mother means so much to you Eve we'll settle here. The day came when those of the first generation that decided not live at the settlement would return to the Time Colony. With a tear in her eye Gladys gave Adam a huge; it would be the last time she would ever be able to make physical contact with him.

CHAPTER 19

As the other Time Craft headed back to the Time Colony Mark and Gladys headed the Omega into the future to monitor the life of Adam and the other settlement members. Mark and Gladys were sad to see the graves of Adam and Eve but the fact they would have other lives to live gave them comfort. As for their offspring's and those other settlement members living among the Cro-Magnons they would adopt many of each other's ways. Eventual one of Adam and Eve's descents would become Pharaoh of Egypt. One of the things handed down through the generations was their belief in an after life.

Although Mark and Gladys had spent many generations in the future they arrived at the Time Colony only moments after the other Time Craft from the settlement. Gladys why don't you take one of the Omega scout crafts and go visit your parents I know they want to hear about their grandson and his descents. What are you going to be doing? I'll meet you there after my debriefing by the President and Planning Committee. At the debriefing there was the question of where every one should be giving the Youth Drink. The President and Committee voted that those who want the Youth Drink must agree not to have any children.

As you know Mark we still have to solve the problem of materializing. I've been thinking about that Mr. President and I think I have the solution. What is it? Send ROD on a time trip long enough so it can evolve to the point able to solve the problem. Sounds like a good idea. OK Mark I'll leave it up to you. ROD I want you to use one of Omega's scout craft and take a time trek back in time until you have evolve to the point you can find a way for us to physically exist in a time other than which we were born. What if I never find a way Captain? Well then when you reach the beginning of the Big Bang come back. ROD had no sooner disappeared heading for the scout craft than suddenly the robot reappeared. Did you forget something? No, Captain I've returned. What? Yes, the more I evolve the greater my ability to be precise. Did you find a way? No Captain. I ordered you if necessary to travel all the way to the beginning of the Big Bang. I did Captain. And you still were unable to find a way? That is right; but there is a way to live in the outside world. How is that?

I discovered a way you could live within those that exist out side. How? I can transfer your conscious being into a person living in real time. What happens to the person? You become that person. Want the person know? No, it takes place before the person is old enough to have a memory. If I consciously exist in someone else what happens to the now me? You would to be placed in suspended animation. Will I remember this life? No, you'll only remember the events in the life you are living except in rare cases such as deja vu. Will Gladys and I recognize one another if we live our lives in the same real time? You would not consciously remember although you would have feeling of deja vu that you've known her before. What would happen if I became someone who already has lived long enough to have a memory? I don't know Captain.

CHAPTER 20

Gladys did you ever think what would happen if someone found out their descents had founded a way to travel through time. I think I'm going to try a little experiment. Mark set course for the year 1950 where in a state mental hospital he found the perfect subject for his experiment by the name of R.C. Terry. R. C. was a patient who had suffered a dilemma in the year 1950 that caused him to go into self induced amnesia. Diagnosed as schizophrenia no one would believe anything as seemly impossible as what Mark would make known to him.

When R.C. Terry awoke that warm spring morning in the state mental hospital just out side of Los Angeles California he couldn't help thinking about the dream he had during the night and how it had seemed so real. However, now he had to start thinking about getting a job and making

a life for his wife and son for today he was going home.

As the years went by R.C. often though of the strange

unexplained things that have occurred in his life. The first

time was in the year 1956 while working in an East Los

Angles Drive-in restaurant. It had been a busy night and he

was taking a rest break when a picture of two cars colliding

appeared in the optical center of his brain. Looking out the

window to see if any one was hurt in the collision he was

startle to find there were no cars. Standing in the window

thinking he was loosing his mind he suddenly saw the two

cars coming down the street. He wanted to run out to the

street and warn them but he figured people would think he

was craze. Then just as he knew it would happen the cars

one making a right turn the other making a left turn at the

same time down the same street collide. It was the first of

many such occurrences and it was a shocker. For months R.C. asked himself how could something so seemly impossible occur. Until then he'd contributed such stories to con-artist. About a month later while driving home from work one night he saw in the visual part of his brain men at work removing trolley cars tracks from a bridge. This was over five miles before the bridge was in view. The awareness that he could foresee his own death sent a cold chill up his spine. Years later his fear almost became a reality for while driving along a road he had never driven before when things began to look familiar and then it came to him; he had dreamed about driving the road the night before. In the dream he was driving along the same road when coming from the opposite direction was a British range rover. As the rover got near it swerved into his side of the road and then he woke up. As the dream unfolded he slowed downed and pulled off the road and just as in the dream when the range rover reached the point in his dream swerved onto his side of the road. At the speed they were going it is unlikely anyone would have survived. Through the years R.C. would have several more experiences of see things occur before they happened. Although he would try to force himself to see things take place before they actual took place he was unable. It was one of those things that just happened.

Not until R.C. was in his old age that he realized the dream he had back in 1950 was more than a dream. It explained why among all the species; Man was the only one able to create that which had never existed before such as clocks, cars, airplanes, etc. It explained why man through the ages had believed himself to be more than just another physical being but part of a greater creator. It didn't take him long to realize he as well as all Humans that put the Human Race above all else was being partners with the time traveling future generation. For they are the greater creator.

The End

www.ingramcontent.com/pod-product-compliance
Lightning Source LLC
Chambersburg PA
CBHW022057170526
45157CB00004B/1388